Brice Sounna Zeuguim
Mama Mouchili

# Breeding manual for commercial laying hens

**Brice Sounna Zeuguim**
**Mama Mouchili**

# Breeding manual for commercial laying hens

## Comparative study of two laying strains in the rearing phase.

**ScienciaScripts**

**Imprint**
Any brand names and product names mentioned in this book are subject to trademark, brand or patent protection and are trademarks or registered trademarks of their respective holders. The use of brand names, product names, common names, trade names, product descriptions etc. even without a particular marking in this work is in no way to be construed to mean that such names may be regarded as unrestricted in respect of trademark and brand protection legislation and could thus be used by anyone.

Cover image: www.ingimage.com

This book is a translation from the original published under ISBN 978-620-6-70972-5.

Publisher:
Sciencia Scripts
is a trademark of
Dodo Books Indian Ocean Ltd. and OmniScriptum S.R.L publishing group

120 High Road, East Finchley, London, N2 9ED, United Kingdom
Str. Armeneasca 28/1, office 1, Chisinau MD-2012, Republic of Moldova, Europe
Printed at: see last page
**ISBN: 978-620-8-21734-1**

# TABLE OF CONTENTS

# DEDICATION

*My parents*

# INTRODUCTION

In sub-Saharan Africa, the demand for table eggs is constantly growing, with local production far below demand. This is due to several factors, including the choice of a strain with high production potential and adapted to our environment. Several strains of imported egg-laying hens are used in African countries, and the literature provides very little information on which strain is best suited to our environment, yet this parameter is very important for local producers who want to increase their yields.

This rearing manual is divided into two parts. The first part is a literature review on rearing techniques for commercial laying hens, in which we describe housing, feeding and sanitary prophylaxis techniques. In the second part, we carried out a comparative study of the growth performance of two commercial laying strains in the rearing phase in West Cameroon, with the main aim of improving knowledge of laying strains adapted to our environment. This work will enable producers to make a better choice of strains for improved performance.

# CHAPTER I REARING COMMERCIAL LAYING HENS

## I-       GENERAL INFORMATION ON COMMERCIAL LAYING HENS

### I.1 Definition of a commercial layer and some examples

Commercial laying hens are selected for economic production. They have high egg production potential (over 300 eggs at 80 weeks) and great hardiness. In tropical Africa, most commercial strains are imported from Europe by countries that have invested substantial financial resources in research to develop their own strains. So, depending on market requirements, there are several categories of egg-laying strains:

- ➢ **Strains that produce red eggs such as Lohmann Brown, Isa Brown, Norvogen Brown, Hy-line Brown...**

- ➢ **Strains that produce white eggs such as Lohmann White, Isa White, Norvogen White...**

- ➢ **Strains are also selected according to egg size.**

### I.2 Some characteristics of commercial laying hens

| DURING THE BREEDING PERIOD | Viability in the rearing period (before laying) between 97% and 98%. |
|---|---|
| | Consumption at 20 weeks between 6.5 Kg and 7.5 Kg |
| | Weight at 20 weeks between 1.5 Kg and 1.7 Kg |
| DURING EGG PRODUCTION | Laying between 18 and 20 weeks |
| | Peak egg-laying between 94% and 96%. |
| | Laying feed consumption between 105 and 120 grams per day |
| | Egg-laying viability between 94% and 96%. |
| | Number of eggs at 80 weeks between 300 and 350 |

### I.3 LAYING HEN BUILDINGS

#### I.3.1 Characteristics of laying hen buildings

Laying hen rearing buildings must meet a number of requirements to ensure their well-being. These requirements include

> **Protecting chickens from the elements (rain, sunshine...);**

> **Protection against disease-carrying rodents and birds;**

> **Building height between 2.5 and 3m;**

> **Position perpendicular to the prevailing wind;**

> **Wall height along the length perpendicular to the prevailing wind from 0.5m to 0.75m and the remainder 2m or 2.5 m screened;**

> **A foot bath at the entrance;**

> **Provide a storage area for feed and eggs;**

> **Easy vehicle access;**

> **Built away from other farms and towns to avoid noise;**

#### I.3.2 The different types of buildings for rearing hens on the ground

There are different types of building for housing hens, depending on their age and the climatic requirements of the rearing environment. There are chick houses and production buildings.

Chick houses are suitable for rearing chicks during their first month. These buildings must be able to regulate the ambient temperature, which must decrease (from 36°C to 20°C) from the first to the sixth week, depending on the chicks' heat requirements. In these buildings, in addition to feeders and drinkers adapted to the chicks, there are also heating systems to regulate the ambient temperature. Examples of heating equipment include electric or gas radiant heaters and wood-fired ovens. These buildings are generally enclosed with a controlled ventilation system.

Production buildings are built to meet the requirements of the hen during the laying period. There are two main types of building:

> **Open buildings are built for natural ventilation, and are positioned perpendicular to the prevailing winds to allow odors to escape and the building to breathe.**

> **The enclosed buildings have an artificial ventilation system. All ambient parameters are monitored.**

**Photo 1**:Open barn

### I.3.3 Housing hens in cages

Housing laying hens in cages has solved the shortage problem. By stacking hens in the cages, we can increase the number of hens per $m^2$ of floor space.

### I.3.3.1 Description of hen cages

The cages are wire-mesh and made of galvanized steel wire, allowing air to circulate. The floor is slightly inclined and extends into an egg-rolling basket that is sufficiently far away and isolated from the hens to prevent them from pricking their eggs. This floor, which must support the weight of the hens, must be rigid, as it will also contribute to the hens' well-being.

Individual pipette drinkers should be placed at the bottom of the cage, not above the feed.

Each hen must occupy a minimum cage space of $450cm^2$ .

At the entrance to the cages, you need a linear feeder that can be fed automatically by a chain system. It can also receive feed manually.

Cages can be stacked in blocks of three or four to form what we call cage batteries. A droppings collection device can be installed between each block. Depending on the degree of automation of the cages, you can have:

> ➤ A manual droppings collection system consisting of fixed plates;

> ➤ A droppings collection system consisting of a long conveyor belt that receives droppings from the hens. This belt allows droppings to be evacuated at the end of the batteries.

### I.3.3.2 Description of two cage battery systems

### I.3.3.2.1 Semi-Californian or close Californian system

In the semi-Californian system, the cages are stacked one above the other, 10 to 20 cm above the ground. The ceilings of the cages are slightly inclined towards the center and covered with protective plates onto which the droppings of the hens on the upper floors fall. The droppings fall into a specially-designed pit.

In this semi-Californian layout, the cages are stacked in the shape of a pyramid.

### I.3.3.2.2 Compact system device

This system has been developed in very cold countries, and provides a higher hen density per $m^2$ . It consists of cages stacked 2 by 2 against each other. Each floor is equipped with a droppings collection system, which can be a conveyor belt.

### I.3.3.3 Advantages of battery housing for hens

The advantages of the cage system are numerous. We have among others:

> ➤ By stacking chickens on top of each other, you can raise a large flock in a small area;

> ➤ When the system is automated, it reduces labor costs compared with a floor-based farming system;

> ➤ As the hens can't come into contact with the eggs, egg breakage is reduced;

> ➤ In this system, the eggs are not in contact with the droppings, so they're cleaner and won't be contaminated by the microbes in the droppings, so they'll have superior microbial quality compared with eggs from hens raised on the ground and laying in nests;

➢ The mortality rate of caged hens is significantly lower than that of free-range hens, simply because they are not in constant contact with their droppings.

➢ Constant dejecta evacuation reduces the parasite load in the building;

➢ The feed consumption of caged hens is much lower than that of free-range hens, and feed wastage is also reduced;

➢ The laying rate of hens raised in cages is higher than that of hens raised on the ground.

**I.3.3.4 Disadvantages of the battery hen housing system**

Raising chickens in cages doesn't only have its advantages. There are also some disadvantages:

➢ Chicken droppings are constantly being removed, and a drying system must be installed to prevent the proliferation of odors;

➢ Setting up this system requires a higher level of financing than for ground-based farming;

➢ In the event of mechanical failure in the automatic collection system, manual feed service and manual egg collection will be very difficult;

➢ Poor presentation of cull hens, which have very poor plumage;

➢ The welfare of the hens is not a priority in this system. The hens are crammed into cages and have no perches to exercise on.

**I.3.4 Equipment for floor hen houses**

The equipment described in this document is for floor rearing on litter. In the case of battery rearing, the rearing equipment includes feeders with drinking, egg collection and droppings collection systems.

For floor rearing, we distinguish between equipment for nursery buildings and equipment for production buildings.

**I.3.4.1 Chick house equipment**

Nursery equipment must be adapted to the age and needs of the birds. There are feeders, drinkers, temperature control equipment and lighting equipment. All this equipment can be manual or automatic, depending on the degree of modernization of your farm.

### I.3.4.1.1 Feeders

There are several different types of chick feeder:

> **Circular trays with a diameter of approx. 40cm;**

> **Adaptable cardboard boxes;**

> **Food chains adapted to chicks.**

### I.3.4.1.2 Chick drinkers

There are also manual drinkers with a water capacity of 4 or 5 liters, and automatic drinkers such as pipettes and siphon drinkers. For manual drinkers, the water service is manual, and you need to check the water level in the drinker each time you fill it. Automatic drinkers, on the other hand, are fed from a cistern via pipes, and the water is automatically renewed in the drinker each time the chicks drink.

As for the number of subjects per trough, we can divide them up as follows:

> **Manual 4 or 5-liter circular drinker: 1 drinker per 100 chicks;**

> **For automatic pipettes, use 1 pipette for 6 or 8 chicks.**

### I.3.4.1.3 Heating equipment for room temperature control

There are several types of heating system for livestock buildings, ranging from modern to traditional. The main heating systems used in tropical Africa are as follows:

> **Wood-fired ovens, also known as brulot. Here, heat is produced by the combustion of wood in the oven. Charcoal can also be used, but it's important to remember that the gas produced by charcoal combustion is carbon monoxide, which binds to hemoglobin in the same way as oxygen, creating a competition between oxygen and carbon monoxide that can asphyxiate chicks. The building must be well ventilated to evacuate the gases produced.**

> **100W incandescent bulbs. These bulbs will give off enough heat to warm the chicks if installed at the right height.**

> **Electric or gas radiant heaters. The advantage here is that the temperature can be easily adjusted.**

### I.3.4.1.4 Bedding

Litter also plays a part in ensuring the animals' comfort, so it must be non-toxic for the chicks. You can use untreated wood shavings or even grass straw.

### I.3.4.2 Equipment for a production building

In production buildings, feeders and drinkers will always be available, but they will have to be adapted to the age of the hens in production and comply with specific breeding standards. In the case of automatic circular drinkers, for example, there should be one for a maximum of 100 hens. There are several types of adult feeder:

> **Linear wooden adult feeder;**

> **Adult siphon feeder, 3 feeders per 100 hens;**

> **Food chain, 5cm linear per hen on the chain.**

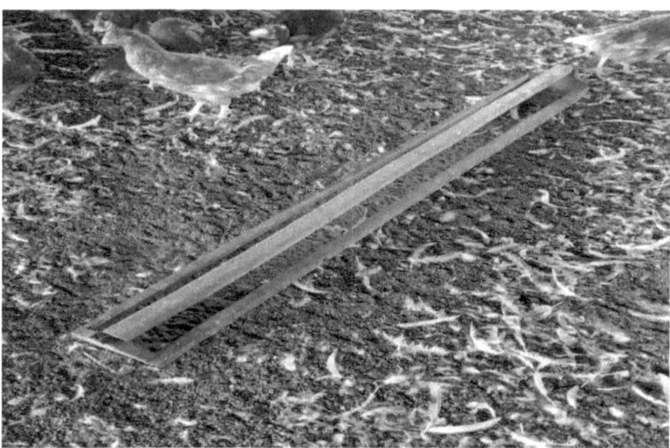

**Photo 2:**Wooden linear feeder

In addition to drinkers and feeders, you'll also need nesting boxes in the production building. The hens will lay their eggs in these nests. Eggs are collected every day, at least twice a day. There should be at least one nest for every six hens. Alongside the nesting boxes, perches can also be provided, as hens like to perch, so a perch is needed in the building to allow them to exercise their natural behavior.

**Photo 3a:**Multi-nest weir

**Photo 3b:** Perches

### I.4 Feeding commercial laying hens

### I.4.1 Types of feed formulation for commercial laying hens

For proper growth, it's important to feed the hens a good-quality feed that meets their nutritional requirements at the right age. In general, two feed formulations will suffice for the first phase of rearing: a starter feed to be fed to chicks up to 8 weeks of age, and a development feed, also known as pullet feed, to be fed until they start laying. When chicks start laying, their needs change, so the feed formulation has to be revised to meet them. Depending on the level of production, you'll have a pre-laying feed served up to 5% of laying (between 19 and 20 weeks), a first phase laying feed served up to 40 weeks, a second phase laying feed served up to 65 weeks and finally a third phase laying feed served from 65 weeks to reformation.

#### I.4.1.1 Starter feed

Starter feed is distributed from the first to the eighth week. At the chick stage, a crumb feed should be served to promote good ingestion without sorting out coarse particles. This feed will have the highest energy and protein content of the entire layer cycle, i.e. an energy content of between 2900 Kcal and 3000 Kcal and a protein content of between 20 and 21%. This ensures rapid chick growth. It should be noted that during this phase, a supply of very fine particles or a very large quantity of coarse elements would lead to selective ingestion of the particles, resulting in an irregular supply of nutrients. In this context, as requirements are not covered, poor growth is to be expected.

<u>**EXAMPLE OF A STARTER FEED FORMULATION**</u>

| Ingredients | | Quantity (in Kg) |
|---|---|---|
| corn | | 600 |
| Wheat bran | | 40 |
| Cotton cake | | 80 |
| Soybean meal | | 100 |
| Peanut meal | | 100 |
| Palm kernel cake | | 10 |
| shell | | 10 |
| Bone powder | | 10 |
| 5% premix | | 50 |
| **TOTAL** | | **1000** |
| Prebiotic additive | | 2Kg/tonne |
| **Nutritional characteristics** | Energy (Kcal) | 2860 |
| | Crude protein (%) | 20,2 |
| | Fat content (%) | 3,9 |
| | Gross cellulose (%) | 6,09 |
| | Lysine (%) | 0,98 |
| | Methionine (%) | 0,38 |
| | Calcium (%) | 1,1 |
| | Phosphorus (%) | 0,5 |
| | | |

### I.4.1.2 Development feed or pullet feed

The switch from starter feed to pullet feed must be made when the weight recommended by the strain standard has been reached. The pullet feed has a lower energy and protein content than the starter feed, to prevent fattening and ensure proper development of the hen's internal organs. This low-nutrient feed is also beneficial in that it helps develop the hen's ingestion capacity.

## <u>EXAMPLE OF CHICKEN FEED FORMULATION</u>

| Ingredients | | Quantity (in Kg) |
|---|---|---|
| corn | | 570 |
| Wheat bran | | 100 |
| Cotton cake | | 60 |
| Soybean meal | | 70 |
| Peanut meal | | 70 |
| Palm kernel cake | | 60 |
| shell | | 10 |
| Bone powder | | 10 |
| 5% premix | | 50 |
| **TOTAL** | | **1000** |
| Prebiotic additive | | 2Kg/tonne |
| **Nutritional characteristics** | Energy (Kcal) | 2720 |
| | Crude protein (%) | 18,2 |
| | Fat content (%) | 3,6 |
| | Gross cellulose (%) | 6,6 |
| | Lysine (%) | 0,86 |
| | Methionine (%) | 0,35 |
| | Calcium (%) | 1,1 |
| | Phosphorus (%) | 0,56 |
| | | |

### I.4.1.3 Pre-foal feed

This feed has a calcium content of around 2%, twice that of pullet feed, and a higher protein content. It should be fed at around $18^{ième}$ weeks, i.e. two weeks before laying. This feed enables the hen to better store calcium to support the upcoming laying period. For precocious pullets, this feed will satisfy the calcium requirements linked to eggshell formation, and for pullets that have been delayed in their growth, it will enable them to catch up, as it contains more nutrients than the development feed.

### EXAMPLE OF PRE-SPONGE FEED FORMULATION

| Ingredients | | Quantity (in Kg) |
|---|---|---|
| corn | | 550 |
| Wheat bran | | 80 |
| Cotton cake | | 80 |
| Soybean meal | | 80 |
| Peanut meal | | 70 |
| Palm kernel cake | | 40 |
| shell | | 40 |
| Bone powder | | 10 |
| 5% premix | | 50 |
| **TOTAL** | | **1000** |
| Prebiotic additive | | 2Kg/tonne |
| **Nutritional characteristics** | Energy (Kcal) | 2669 |
| | Crude protein (%) | 18,7 |
| | Fat content (%) | 3,6 |
| | Gross cellulose (%) | 6,3 |
| | Lysine (%) | 0,9 |
| | Methionine (%) | 0,36 |
| | Calcium (%) | 2,25 |
| | Phosphorus (%) | 0,53 |
| | | |

### I.4.1.4 Egg-laying feed

The egg-laying feed will meet the hen's needs during the egg-laying period. During this phase, the hen's calcium requirements will increase from 4% to 4.4% as the hens age. The egg-laying feed must cover the hen's maintenance, growth and egg-laying requirements.

Pigments such as xanthophylls are also important in the food to reinforce the yolk color.

## EXAMPLES OF EGG-LAYING FEED FORMULATIONS

| Ingredients | | Quantity (in Kg) | | |
|---|---|---|---|---|
| | | Laying 1 | Laying 2 | Laying 3 |
| corn | | 550 | 560 | 570 |
| Wheat bran | | 55 | 50 | 50 |
| Cotton cake | | 80 | 70 | 70 |
| Soybean meal | | 80 | 80 | 70 |
| Peanut meal | | 65 | 65 | 60 |
| Palm kernel cake | | 30 | 30 | 30 |
| shell | | 80 | 85 | 85 |
| Bone powder | | 10 | 10 | 15 |
| 5% premix | | 50 | 50 | 50 |
| **TOTAL** | | **1000** | **1000** | **1000** |
| Prebiotic additive | | 2Kg/tonne | 2Kg/tonne | 2Kg/tonne |
| **Nutritional characteristics** | Energy (Kcal) | 2606 | 2608 | 2604 |
| | Crude protein (%) | 18 | 17,5 | 17 |
| | Fat content (%) | 3,4 | 3,4 | 3,4 |
| | Gross cellulose (%) | 5,8 | 5,7 | 5,7 |
| | Lysine (%) | 0,87 | 0,85 | 0,81 |
| | Methionine (%) | 0,36 | 0,34 | 0,33 |
| | Calcium (%) | 3,76 | 3,95 | 4,1 |
| | Phosphorus (%) | 0,5 | 0,5 | 0,55 |
| | | | | |

### I.4.2 Laying hen requirements

The table below shows the requirements of laying hens according to the evolution of their physiological condition.

| Nutritive element | Start-up phase | Chicken phase | Pre-axle | Laying 1 | Laying 2 | Laying 3 |
|---|---|---|---|---|---|---|
| Metabolizable energy (Kcal) | 2900 - 3000 | 2750 - 2800 | 2750 - 2800 | 2700 - 2750 | 2700 - 2750 | 2700 - 2750 |
| Crude protein (%) | 20 - 21 | 16 - 17 | 16 - 17 | 18,7 | 18,2 | 18,0 |
| Fat content (%) | 3,5 - 5 | 2,5 - 4 | 2,5 - 4,5 | 2,5 - 5,5 | 2 - 4,5 | 1,5 - 3,5 |
| Fiber (%) | 2,5 - 3 | 2,5 - 6,5 | 3,5 - 5,5 | 3,5 - 6 | 3,5 - 6,5 | 3,5 - 7 |
| Calcium (%) | 1 - 1,1 | 0,9 - 1 | 2,2 - 2,5 | 4,1 | 4,2 | 4,4 |
| Phosphorus (%) | 0,45 - 0,5 | 0,35 - 0,4 | 0,42- 0,45 | 0,42 | 0,4 | 0,38 |
| Lysine (%) | 1 | 0,67 | 0,8 | 0,85 | 0,85 | 0,8 |
| Methionine (%) | 0,48 | 0,34 | 0,36 | 0,46 | 0,43 | 0,38 |
|  |  |  |  |  |  |  |

The table above shows the requirements of laying hens at each stage of development. These requirements can be used to formulate different rations to meet the laying hen's needs.

Feed formulation involves combining a number of raw materials and feed supplements to meet the animal's needs for various nutrients (energy, protein, fat, fiber, amino acids, minerals and vitamins).

Knowledge of the bromatological value of the various ingredients available, and of their limits of use, is essential for good food formulation.

As well as nutrients, hens also need water for proper digestion. As a general rule, hens consume twice as much water as food. But this consumption is strongly influenced by ambient temperature.

At an ambient temperature of around 20°C, water consumption will be between 1.8 and 2 times that of feed, and around 2.1 times during the laying period.

Water consumption will increase with temperature, so the hotter it is, the more water the hen will consume. This consumption will also vary with water temperature, so cooler water will be consumed more by the hens.

## I.5 CHICKEN DISEASES AND MEDICAL PROPHYLAXIS

### I.5.1 Some diseases encountered in laying hen farming

Diseases affecting laying hens fall into several categories: viral diseases, bacterial diseases, parasitic diseases and digestive diseases. In addition to diseases caused by microbes, there are also those caused by mineral and vitamin deficiencies.

#### I.5.1.1 Viral diseases

Viral diseases, as their names suggest, are caused by a virus, and generally result in significant economic losses. The main ones are as follows:

> ➢ Marek's disease, manifested by digestive disorders, weight loss, eye discoloration and blindness, and nervous symptoms such as torticollis.

> ➢ Newcastle disease or avian pseudo-plague, which manifests itself through respiratory disorders and nervous symptoms such as paralexia, torticollis and others.

> ➢ Infectious bronchitis is also a respiratory disease, with affected individuals suffering from respiratory problems, and adult egg-layers showing white eggshells, often with blood spots;

> ➢ Gumboro disease, also known as infectious bursitis, is a virus that attacks the bursa of Fabricius, causing intramuscular hemorrhages (petechiae in the leg muscles);

> ➢ Smallpox, or avian diphtheria, manifests itself in cutaneous and oculonasal forms, with lesions on the head and crest.

### I.5.1.1.1 Marek's disease

| Disease | **Marek's disease** | |
|---|---|---|
| Synonymous names | Fatty liver disease, chicken paralysis, avian herpes virosis. | |
| definition | Infectious, contagious, transmissible avian viral disease; tumoral, lymphoproliferative disease; damage to the lymphopoietic immune system. | |
| Epidemiology | Age:young (6 to 7/ 7 to 16 wks) weeks adult:> 16wks<br>Species: more pathogenic in hens; less pathogenic in turkeys | |
| Etiology (causes) | Pathogen | Herpes virus: pathogenic in intracellular environment, resistant to 20°C for 8 months |
| | Contaminating materials | Saliva, nasal secretions, feathers (desquamation of feather follicles) |
| | Penetration routes | Oral, digestive; air, respiratory |
| | | |
| symptoms | Incubation: a few weeks to a few months;<br>Progression from acute visceral form to chronic nervous form, superacute form: 4 weeks;<br>100% mortality without symptoms.<br><u>Acute visceral form</u>:6 to 7 wks/7 to 16 wks; rapid evolution (2 to 5 days); pale crest and barbels; "gooseflesh"-like skin.<br><u>Chronic (nervous) form</u>:> 16 wks<br>Slow evolution (1 to 3 weeks); progressive neck paralysis (cervical nerve), torticollis and ruffle atomia, drooping wings, wide-spread legs, iris discoloration (glass eye or fish eye), egg-laying drop. | |
| diagnosis | <u>Clinical diagnosis</u>: glass eye, goosebumps on the skin, tumors (T lymphocytic infiltration deforming the affected organ);<br><u>Paraclinical diagnosis</u>: Histological, by identifying the predominant lymphocyte type (predominantly LT infiltration); Serological, by ELISA testing for virus detection.<br><u>Differential diagnosis</u>: avian leukosis (visceral form); avian smallpox (cutaneous form). | |
| treatment | No treatment | |
| prophylaxis | Hatchery vaccination: 3 forms of vaccine available (freeze-dried HVT serotype 3, frozen HVT serotype 3 and attenuated frozen serotype 1). | |

### I.5.1.1.2 NEWCASTLE disease

| Disease | | New Castle disease |
|---|---|---|
| Synonymous names | | Avian pseudo-plague; paramyxovirosis |
| definition | | Viral, infectious, contagious disease |
| Epidemiology | | All avian species; all ages, all seasons, |
| Etiology (causes) | Pathogen | Paramyxovirus ( Paramyxoviriidae); enveloped RNA virus; 2 genera: Metapneumovirus (infectious rhinotracheitis and big-head syndrome virus) and Avulavirus (New Castle virus). |
| | Contaminating materials | Excretion:jetage (nasal mucus) and droppings |
| | Penetration routes | Oral and respiratory route |
| | character | Resistant for 2 to 3 years in the cold; destroyed by heat in a few minutes at 60°C and in a few seconds at 100°C. |
| symptoms | | • **Incubation:** 2 **to** 18 days or 4 to 15 days depending on the pathogen strain<br>• **Symptoms evolve in 3 phases:**<br>-Invasion phase (septicemic phase): inappetence, depression, leading to death within 24 hours;<br>-state phase (acute phase): 24 to 48 hours, nervous disorders (hanging wings, torticollis); respiratory disorders (open beak, mucous in the trachea, noisy breathing); ocular disorders: coryza syndrome (moist, swollen eyes); digestive disorders: greenish diarrhea.<br>terminal phase: respiratory + nervous + digestive disorders associated with symptoms of coughing, inappetence and drop in egg-laying. |
| diagnosis | | -Epidemiological diagnosis: high mortality and contagiousness, all ages and seasons<br>- Clinical diagnosis: very rapid evolution (24 hours) + high mortality; lesion of the digestive tract.<br>-Paraclinical diagnosis: laboratory testing for virus identification (hemagglutinin, neutralizing antibodies, subtypes) using serology: IFD (immunofluorescence direct) + IHA (inhibition of hemagglutination) + HAP (passive hemagglutination) + ELISA. |
| treatment | | Antibiotic therapy for bacterial complications only |
| prophylaxis | | Medical prophylaxis:vaccination;<br>Sanitary prophylaxis: hygiene and disinfection of breeding buildings and equipment, sanitary vacuum after each batch. |
| | | |

### I.5.1.1.3 Infectious bronchitis

| Disease | | Infectious bronchitis |
|---|---|---|
| Synonymous names | | Corona virosis of hens; Tracheobronchial syndrome; Syndrome Bronchoovaritis; Broncho-salpingitis syndrome. |
| definition | | Viral disease Infectious Contagious-Inoculable with tropism Respiratory + renal + genital. |
| Epidemiology | | Species: Gallus is more sensitive<br>Age: young people are more sensitive |
| Etiology (causes) | Pathogen | Coronavirus - single-stranded RNA virus, from 80 to 160 nm; intracellular multiplication - a dozen serotypes;<br>In vivo culture (tracheal and bronchial epithelial cells) + in ovo (9-10 day-old embryonated egg).<br>-Characteristics: Withstands 1 month in droppings and droppings in the outdoor environment + a few weeks in litter.<br>-Sensitive:formalin + UV + Incubator (37°Cfor12h) - Disinfectants. |
| | Contaminating materials | Nasal + oral + ocular secretions and droppings |
| | Penetration routes | Respiratory: Trachea horizontal transmission: Direct and Indirect. |
| | | |
| symptoms | | **Incubation**: 18 to 36h (very short)<br>Symptoms vary according to age and tropism (3 tropisms: respiratory, genital and renal).<br>**-General symptoms**:Depression and anorexia.<br> **-Respiratory symptoms**: Dyspnea, cough, rales, sneezing, bloody sputum, coryza syndrome, conjunctivitis and sinusitis. Progression to CKD or death in 2 to 3 weeks, or recovery.<br>**Genital symptoms**:Sudden drop in egg laying (10-50%) => Poor layers => deformed eggs (40%) + Discolored shell + Thin or absent shell + Granular or rough shell + Albumen less viscous (too liquid).<br>**-Kidney symptoms**:<br>Depression, intense thirst, wet droppings.<br>**-Lesions**: respiratory, genital and renal<br>Evolution:Death. |
| diagnosis | | Paraclinical diagnosis:<br>-Histology: evidence of virus in tracheal epithelial cells<br>-Virology: virus detection by immunofluorescence.<br>-Serology: ELISA test, |
| treatment | | No treatment; antibiotic therapy is used only to prevent evolution towards a<br>RCM |
| prophylaxis | | -sanitary prophylaxis: hygiene and disinfection<br>-medical prophylaxis: live attenuated vaccine, nebulization |

### I.5.1.1.4  Avian influenza

| Disease | | Avian influenza |
|---|---|---|
| Synonymous names | | Avian influenza; Avian plague |
| definition | | Highly contagious, inoculable infectious viral disease. Zoonosis.<br>Respiratory + Digestive + Nervous Tropism |
| Epidemiology | | Domestic, wild and exotic birds |
| Etiology (causes) | Pathogen | The pathogen is a helically symmetrical ribovirus or enveloped RNA virus of the Orthomyxoviridae family, of the "Influenza" genus. Presents 3 antigenic types (influenza virus type A in birds; influenza virus types B and C in humans). 16 types of hemagglutinin + 9 types of neuraminidase. Highly pathogenic, multiplying easily in embryonated eggs. |
| | contaminants | Source of contamination:Wild birds (migratory), especially palmipeds. The contaminants are droppings and corpses. |
| | Penetration routes | By respiratory and digestive tracts |
| | | |
| symptoms | | -Pathogenic influenza (A strain):<br>Respiratory distress + lacrimation + sinusitis + head edema + cyanosis of crest and barbels + diarrhea=> 100% mortality.<br>-Moderately pathogenic influenza (B strain):<br>Respiratory disorders + Aerosacculitis + sudden drop in or cessation of egg-laying; 50-70% Mortality + High morbidity.<br>Non-pathogenic influenza (C strain):<br>Slight respiratory and digestive problems - slight drop in egg production. |
| diagnosis | | Paraclinical diagnosis<br>Laboratory testing for virus identification (hemagglutinin, neutralizing Ac, subtypes) using serology: HIA (Hemagglutination Inhibition) + ELISA. |
| treatment | | No treatment, just antibiotic therapy for bacterial complications |
| prophylaxis | | Disinfection of buildings and equipment; quarantine; destruction of contaminated herds; destruction of bedding by fire. |

### I.5.1.1.5 Avian encephalomyelitis

| Disease | | Avian infectious encephalomyelitis (EA) |
|---|---|---|
| Synonymous names | | Epidemic chicken tremor disease - Piano player disease - Picornavirosis. |
| definition | | Viral disease - Infectious - Contagious - Inoculable - Tropism; Enterotropic + Neurotropic. |
| Epidemiology | | Birds: Chicken-Turkey-Faisan-Quail-Perdrix-Pintade. Age: Very young - Older. |
| Etiology (causes) | Pathogen | Pathogen: Single-stranded RNA virus - small size (20 to 30 nm) - Picornaviridae family - similar to hepatitis virus (genus Hepatovirus) - no antigenic variants (pathogenicity varies according to strain). Characters: Resistance: several weeks to several months in the external environment. Sensitive:Usual disinfectants. |
| | Contaminating materials | Droppings |
| | Penetration routes | Oral (digestive) Transmission: -Horizontal: possible (eggs hatching) -Vertical: especially (contaminated eggs) |
| | | |
| symptoms | | Incubation: -1 to 7 days after vertical transmission -4 to 14 days after horizontal transmission. **In chicks (>3sem / >6sem):** 10d: Sudden death without symptoms (contaminated eggs) 15j:Progressive muscular ataxia (progressive motor incoordination with tremor) => head tremor => neck tremor => youngster sitting on tibio-metatarsal joint in "piano player" position => flaccid paralysis => no more movement => death by inanition -Progression: Death or recovery (with growth retardation + Cataracts + Slight paralysis). Morbidity rate: 40-50 -Mortality rate: 25 to 50% (depending on strain + virulence). **In breeding females:** Decreased fecundity and hatchability + Egg-laying drop + Abatement + Decreased CI + Absence of nervous symptoms + Cataracts possible. Egg-laying drop: At the start of egg-laying (slight drop in egg-laying + slow, abnormal return) or during egg-laying (sudden drop in egg-laying + rapid, abnormal return). normal). |
| diagnosis | | Clinical diagnosis: Tremor and ataxia within 3 weeks. Paraclinical diagnosis: -Histology: Nervous tissue + Proventriculus + Gizzard + Pancreas. -Virology: Embryo culture => Immuno-Fluorescence. -Serology: neutralizing Ac on embryonated egg. Diagnosis |
| treatment | | No treatment - Vitamin complex (vitB) - Antibiotic therapy for bacterial complications only. |
| prophylaxis | | -Eliminating sick birds -Vaccinate breeding females before laying with a live attenuated vaccine administered orally in drinking water. |

### I.5.1.1.6 Gumboro disease

| Disease | | Gumboro disease |
|---|---|---|
| Synonymous names | | Infectious bursal disease - IBDV (Infectious Bursal Desease Virus) - Burna virose. |
| definition | | Viral disease - Contagious - Inoculable - Immunosuppressive - Tropism for the bursa of Fabricius (lymphoid organ) |
| Epidemiology | | Species: Chicken mainly - Turkey and Duck (subclinical form) Age: Young (2 to 7 weeks) |
| Etiology (causes) | Pathogen | **Pathogen:** Birna virus - Very stable - Non-enveloped - 60 nm in diameter - Affinity for the bursa of Fabricius => Destruction of B lymphocytes => no immune response. **Characters:** Resistance: >4 months in litter - A few hours at 40°C and 5h at 56°C - Ether - Chloroform - Quaternary ammoniums. Sensitive: 10 min in 5% formalin and 30 min in 7% formalin - Hot water. |
| | Contaminating materials | Source of contamination: Parasitic worms - sick birds. Contaminating materials:Droppings. |
| | Penetration routes | Route of entry: Digestive only. Transmission: Horizontal only: (Direct through soiled litter + Indirect through soiled livestock equipment). |
| symptoms | | -Acute form:2-6 wk Incubation: 2 to 3 days => Duration: 7 days Sudden anemia - Decreased IC - Increased water intake (fever) => Watery, whitish diarrhea => Dehydration + Ruffled feathers. Trembling - Wobbly gait => Abatement - Prostration. Evolution: Death (2 to 4 days) or Recovery (5 to 7 days) with growth retardation. -Subclinical form: >6sem Economic losses: Decreased weight + Increased CI => Significant mortality Evolution: CKD (Chronic Respiratory Diseases). |
| diagnosis | | Epidemiological diagnosis: Age. Clinical diagnosis: Pathognomonic lesions. Paraclinical diagnosis: Histology (BF necrosis lesions) + Immunology (ELISA). Differential diagnosis: -Cold mortality -New Castle disease (petechiae + diarrhea) -Coccidiosis (Hemorrhagic diarrhea). |
| treatment | | No treatment |
| prophylaxis | | Sanitary prophylaxis: Hygiene and disinfection: -Good disinfection: hot water at high pressure - Formol. -Good insect control: burning litter 15-day sanitary vacuum. Medical prophylaxis: Vaccination Day 1, Day 7, Day 14 and Day 21 |

## I.5.1.1.7 Fowl pox

| Disease | | Fowl pox |
|---|---|---|
| Synonymous names | | Avian diphtheria - Contagious epithelioma - Pox virose. |
| definition | | Viral disease - Contagious - Inoculable. |
| Epidemiology | | All birds: Chicken - Turkey - Pigeon - Peacock - Quail - Duck. All ages: especially at the end of production. All seasons |
| Etiology (causes) | Pathogen | Pathogen:<br>Pox virus - DNA virus - 250 nm in diameter - cultivated in ovo by intradermal inoculation - multiplies in an egg containing an embryo (on chorio-allantoic membrane), causing characteristic patches of cell necrosis 1 to 2 mm in diameter = known as "pock".<br>Characters:<br>-Resistant: several years in scales - 9 days in formalin.<br>-Sensitive:Usual disinfectants - 30mn at 50°C. |
| | Contaminating materials | Source of contamination:Healthy carriers, Insects. Contaminating materials: Dander - Dust - Feather debris. |
| | Penetration routes | Entry routes:Skin penetration (biting - mosquito bite - IA) + Mucous membrane penetration<br>Transmission: horizontal and vertical routes are non-existent; only through skin and mucous membrane penetration.<br>Pathogenesis:<br>Penetration => intracellular multiplication => intracytoplasmic inclusions=>opharyngeal mucosa + tracheal mucosa. |
| | | |
| symptoms | | Incubation:4 to 14 days<br>**Cutaneous form:** frequent<br>Nodules on non-feathered parts of the body (cephalic appendages, eye rims, beak corners, mumps, leg scutes and fingers) in the form of: whitish papules => whitish pustules => yellowish vesicles => grayish to brownish crusts => detach after 4 weeks => healing.<br>Reduced weight + Reduced fertility + Egg-laying drop.<br>**Diphtheroid form:**Nodules in ocular and nasal mucosa - Nodules in digestive tract and respiratory tract.<br>Reduced weight + Reduced fertility + Egg-laying drop.<br>Mixed or mucocutaneous form:<br>Both forms can occur at the same time (especially in turkeys). A form known as "variola coryza":<br>Birds show all the signs of ordinary coryza =><br>Evolution: towards spontaneous recovery or bacterial superinfection. |

| diagnosis | -Clinical diagnosis:<br>Symptoms and pathognomonic lesions.<br>-Paraclinical diagnosis:<br>　　* **Virology:** cell culture + PCR.<br>　　* **Serology:** serum neutralization test.<br>　　* **Histology:** Proliferation and hyperplasia of epithelial cells of the epidermis and mucosa, with intracytoplasmic eosinophilic inclusions. |
| --- | --- |
| treatment | -Antigenotherapy= 1/10 ml suspension in the dermis of the barb.<br>-Antibiotic therapy to prevent bacterial superinfections.<br>-Local treatment with iodinated glycerine, silver nitrate or methylene blue solutions. |
| prophylaxis | Hygiene + insect control + attenuated vaccines. |
|  |  |

### I.5.1.1.8　Infectious syndrome of the large head

| Disease | | Infectious big-head syndrome |
| --- | --- | --- |
| Synonymous names | | Avian pneumovirosis; Avian paramyxovirosis |
| definition | | Viral disease, infectious, pneumovirus, respiratory tropism. |
| Epidemiology | | Especially chicken and guinea fowl, all ages. |
| Etiology (causes) | Pathogen | Pathogen:<br>Pneumo virus (paramyxoviridae). This is a Paramyxovirus (paramyxoviridae) - enveloped RNA virus - 2 genera:Genus Metapneumovirus (Infectious Rhinotracheitis and Infectious Large Head Syndrome viruses) and Genus Avula virus (New Castle virus). |
| | Contaminating materials | Nasal and tear discharge |
| | Penetration routes | Transmission:<br>Horizontal direct transmission by:<br>- airborne (inhalation)<br>-ocular and oral routes possible (via the lacrimal duct or cleft palate) |
| | | |
| symptoms | | **Chicken and guinea fowl (young):**<br>Weak rales - nasal and ocular discharge - swelling: of the eyelids in the periorbital region + of the infraorbital sinuses + of the lower mandible + sometimes the nape of the neck (swelling of the head => "big head" appearance).<br>Drowsiness - dehydration - nervous disorders (loss of balance + torticollis) - mortality rate 10%.<br>**Chicken and guinea fowl (adults):**<br>Laying drop ranging from 5% to 30%, which can affect the entire flock in 2 to 3 weeks. |

| diagnosis | **Serological examination**:<br>-Purpose: virus detection<br>-Virus isolation=<br>-From tracheal and ocular swabs (first 3 days) by inoculation into cell cultures.<br>   -From 2 blood samples (first 2 days of infection, then 3 to 4 weeks after infection), using the sero-neutralization and ELISA techniques (ELISA is less effective in turkeys).<br>**Histological examination**:<br>-Purpose = to identify the virus to be used:<br>-Respiratory system damage<br>-Intracytoplasmic inclusions in ciliated epithelial cells of the nasal turbinates and larynx.<br>**Bacteriological examination**:<br>Aim:to identify the complicating germ + establish a differential diagnosis. |
|---|---|
| treatment | -No antiviral treatment.<br>-Antibiotic therapy for bacterial complications. |
| prophylaxis | **Sanitary prophylaxis**:<br>- Strict compliance with hygiene standards (proper disinfection),<br>- Crawl space,<br>- Quarantine,<br>- Single-band rearing,<br>- Reasonable density.<br>**Medical prophylaxis:**<br>- Live virus vaccine: primary vaccination at start of rearing, first booster before laying, second booster during production.<br>- Risk of interference with adenovirus in turkeys, hence vaccination against turkey hemorrhagic enteritis. |
|  |  |

### I.5.1.1.9 Egg-laying drop syndrome

| Disease | | Egg-laying drop syndrome |
|---|---|---|
| Synonymous names | | EDS = egg drop syndrome; avian adenovirus |
| definition | | Viral disease - Infectious - Inoculable-Tropism+ respiratory<br>Hepatic + Splenic + Pancreatic. |
| Epidemiology | | Laying hens - Goose - Wild birds |
| Etiology (causes) | Pathogen | Pathogen:<br>*Adenovirus, Family Adenoviridae=> 4genres:<br>-Mastadeno virus (mammals),<br>-Aviadeno virus (birds) => Group I avian adenoviruses,<br>-Siadeno virus (birds) => Group II avian adenoviruses,<br>-Atadeno virus (birds) => Group III avian adenoviruses.<br>Responsible for egg-laying syndrome in hens, or EDS. |
| | Contaminating materials | All excretions, especially droppings. |
| | Penetration routes | Respiratory and digestive routes; Transmission Vertical (in the egg or female genital tract) or Horizontal (direct and indirect), depending on the virus. |
| | | |
| symptoms | | 50% drop in egg-laying persists for 1 to 3 months; fragile, small eggs; thin, rough, soft, discolored shell => Evolution: Recovery + return of egg-laying curve to normal (same as avian infectious encephalomyelitis). |
| diagnosis | | -Virus isolation and identification.<br>-Cytopathogenic effect (insufficient to make a definitive diagnosis).<br>-E.L.I.S.A.; Complement fixation; Seroneutralization; Hemagglutination inhibition; Double agar immunodiffusion. |
| treatment | | No treatment |
| prophylaxis | | - Strict hygiene measures (in relation to brooding).<br>-Inactivated adjuvanted vaccine: 14 to 18 weeks for laying hens |
| | | |

## I.5.1.2 Bacterial diseases

These illnesses are caused by bacteria such as Salmonella, E. coli and many other microbes that cause digestive problems. In general, the following symptoms are observed:

- ➤ Dead hens rapidly turn black a few hours after death as a result of intense bacterial activity in the digestive tract;

- ➤ Reduced feed consumption;

- ➤ Whitish diarrhea in cases of salmonellosis;

- ➤ At autopsy, a thin, whitish membrane is observed on the viscera in cases of colibacillosis.

The following tables describe the main digestive diseases.

## I.5.1.2.1 Avian colibacillosis

| Disease | | Avian colibacillosis |
|---|---|---|
| Synonymous names | | By disease entity |
| definition | | Bacterial, infectious, contagious, inoculable disease |
| Epidemiology | | All birds (especially chicken and turkey); all ages |
| Etiology (causes) | Pathogen | Pathogen:<br>*Family Enterobacteriacea, genus Escherichia coli, Bacillus, G-, mobile or immobile,<br>*Escherichia coli serotypes: several serotypes (most pathogenic: O1:K1, O2:K1, O78:K80)<br>E.coli is species-specific (e.g. O86 duck)<br>-E.coli is an immunosuppressant<br>Characters:<br>-Resistance: several days in ambient environment (soil and water)<br>-Sensitive: usual disinfectants. |
| | contaminants | Source of contamination:Healthy carriers (permanent shedders of contaminants > 109 germs), contaminated eggs (before, during and after laying).<br>contaminant: droppings |
| | Penetration routes | Routes of entry: oral, respiratory, genital Transmission:<br>-Mainly horizontal (faecal contamination, dust, feathers, water, food, soil, soiled eggshells)<br>-Vertical rare (ovarian cluster, oviduct affected)<br>Predisposing factors:Stress, intercurrent diseases (viral, bacterial, parasitic) |
| | | |
| symptoms | | -Mortality in the egg (embryonic) => drop in hatchability - Just after hatching => limp chick, swollen abdomen, swollen, moist navel;<br>-white diarrhea |

|  | -fibrin deposits (a thin layer of whitish membrane on the liver, heart and abdomen),<br>egg-laying drop,<br>-Sudden death. |
|---|---|
| diagnosis | D. Clinical: fibrin deposits in the liver, heart and abdomen;<br>D. paraclinical: Search for E. coli in droppings, organs (liver, kidneys, oocyte) Isolation and identification of pathogenic strains |
| treatment | Antibiogram to define appropriate treatment (resistant to tetracyclines, streptomycin, chloramphenicol)<br>-Oral: sulfonamides (trimethoprim => sulfaquinoxaline); quinolones (flumequine => Enrofloxacin) in drinking water<br>In combination with macrolides (strepto + spiramycin or strepto + tylosin) for concomitant pathologies |
| prophylaxis | -Chemoprevention and vaccination (very delicate)<br>-Rigorous hygiene of building, equipment and personnel<br>-Sanitary vacuum between each batch (hatching and rearing)<br>-Avoid stress by administering vitamins |
|  |  |

## I.5.1.2.2  Avian Mycoplasmas

| Disease | | Avian mycoplasma |
|---|---|---|
| Synonymous names | | Depending on whether the pathogen stands alone = "Mycoplasmoses", or is associated with other infectious agents = "MRC". |
| definition | | Bacterial, infectious, contagious, inoculable, cosmopolitan disease, |
| Epidemiology | | -All birds (especially chicken, turkey, duck, goose, pigeon, guinea fowl, quail, pheasant, partridge)<br>-Age: all ages |
| Etiology (causes) | Pathogen | Pathogen:<br>*Class Mollicutes, Order Mycoplasmatales, Family Mycoplasmatacea (sterol-dependent) and Family Acholeplasmatacea (sterol-independent), Genus Mycoplasma<br>*Pseudomycelial in appearance, coccoid or pyriform bacteria, branched or star-shaped, filamentous or spiral, motile or immobile, no Gram staining,<br>*Mycoplasma species: Mycoplasma gallisepticum, M. synoviae, M. meleagridis, M. iowae) other species. Mycoplasma is species-specific (e.g. M. meleagridis in turkeys) and clinically specific (e.g. M. synoviae in joints).<br>-Mycoplasma is an immunosuppressant<br>-Characteristics: Withstands a few days to a few weeks in ambient conditions. |
|  | Contaminating materials | -Source of contamination: healthy carriers (permanent excretors of contaminants)<br>-Contaminant: sputum; |

| | | |
|---|---|---|
| | Penetration routes | -Routes of entry:respiratory and genital - Transmission:Horizontal (sputum, dust, litter); Vertical (damage to ovarian cluster, oviduct via hematogenous route or contact with infected air sacs) |
| | | |
| symptoms | | Mycoplasma is responsible for several pathological entities:<br>➤ **Mycoplasma gallisepticum (in hens, turkeys, ducks and partridges):**<br>-growth retardation, increased IC, reduced hatchability and egg production, stillbirths.<br>-Respiratory disorders (coryza, sneezing, jetage, dyspnea, prostration with open beak; in turkeys, uni or bilateral suborbital sinusitis)<br>-Consequences => CKD (chicken) and infectious sinusitis (turkey)<br>➤ **Mycoplasma synoviae: (In hens, turkeys and guinea fowl:**<br>-stunted growth, weakness, pale crests and barbels<br>-less severe upper respiratory disorders (coryza, rales, sinusitis)<br>locomotor disorders (enlarged or swollen joints on legs and/or wings; blisters on the wishbone) aggravated by joint-tropic germs (reovirus) |
| diagnosis | | D. clinical:<br>Symptoms and Lesions (according to causal agent and animal species)<br>D. paraclinical:<br>Mycoplasma testing |
| treatment | | -Antibiogram to determine appropriate treatment<br>Anti-infectives: macrolides (spiramycin, tylosin, tilmicosin), tetracyclines (chlortetracycline, oxytetracycline), aminoglycosides (kanamycin, gentamycin), quinolones (flumequine => Enrofloxacin) |
| prophylaxis | | -Chemoprevention and vaccination (very delicate)<br>-Prevent all forms of infection through rigorous hygiene of farm buildings, equipment and personnel.<br>-Sanitary vacuum between each batch (hatching and rearing)<br>-Avoid all forms of stress |
| | | |

### I.5.1.2.3 Avian salmonellosis

| Disease | | Avian Salmonellosis |
|---|---|---|
| Synonymous names | | Fecal peril - Bronzed liver disease - Blue ridge disease Pullorosis - Typhosis |
| definition | | Bacterial, infectious, contagious, inoculable disease, A digestive and genital tropism with extremely variable organic disorders |
| Epidemiology | | Bird species: All birds |
| Etiology (causes) | Pathogen | Pathogen: Enterobacteriacea family, Salmonella genus, 2 major species: <br> -Salmonella Bongori (rare) <br> -Salmonella choleraesuis => Salmonella enteritica (frequent) => 7 subspecies of Salmonella enteritica including 2 S/E of Salmonella enteritica (more important): <br> *Salmonella enteritica arizona => poultry only <br> *Salmonella enteritica enterica => man + poultry + mammals => several serovars: <br> Salmonella gallinarum-pullorum (SGP) => Pullorosis and Typhosis in birds; <br> Salmonella Typhi => Typhus (typhoid fever) in humans; <br> Salmonella Paratyphi => paratyphoid in humans <br> Salmonella Typhimurium => TIA (zoonosis) <br> **Characters:** <br> **-Resistant to:**cold - salting - smoking - desiccation - 12 days on eggs incubated at 25°C - 2 years in droppings - 9 months in soil and mud (away from sun and heat). <br> **-Sensitive:**UV (sun) - Heat - formalin 10% - common disinfectants. |
| | contaminants | **Source of contamination:** Healthy carriers (permanent excretors = 1 billion germs/gr faeces), Soiled eggs (before, during and after laying), Insects. <br> **Contaminating materials:** Especially droppings, excretions, down or feathers. |
| | Penetration routes | **Routes of entry:**Oral - respiratory (hatching) - genital (coitus) <br> **Transmission:** Horizontal + Vertical <br> **Mortality rate:** <br> -50% => pullorosis <br> -20% => typhosis <br> -20% to 30% => breeding stock |
| | | |
| symptoms | | Shell mortality: <br> 6$^{ème}$ to 15$^{ème}$ days of incubation. <br> ****Pullorosis:** <br> **F. suraigüe:**septic Mortality at 5$^{ème}$ and 15$^{ème}$ d of life. <br> **Acute f.:** chicken 1-2 weeks old |

| | |
|---|---|
| | Sad; frigid; anorexic; hanging wings (weakened); white, chalky, sticky diarrhea, clogging the cloaca as it dries; respiratory distress possible => progresses to death.<br>**-F. chronic**: chicken from 3 to 6 weeks<br>Poor general condition, locomotor disorders (deviated legs), torticollis.<br>****Typhosis:**<br>High-pitched F:<br>Sudden death or depression, fever, cyanosis of the crest "Blue crest disease", death within a few days<br>-F.aigue:<br>Prostration, respiratory disorders (rales, stinky discharge), yellow or green foul-smelling liquid diarrhea streaked with blood that sticks to feathers. Occasional nervous disorders (staggering).<br>-F.chronic:<br>Poor general condition, locomotor disorders, genital disorders (egg-laying drop, blood-stained eggs, eggs without shell, reduced hatchability). |
| diagnosis | D. clinical:Symptoms and lesions<br>D. paraclinical:<br>-Youth: liver + spleen<br>-Adults:liver + spleen + blood<br>D. differential:<br>-Shell mortality:<br>Aspergillosis (black spots at candling) Brooding technical errors |
| treatment | -Oral route: quinolones (flumequine, oxolinic acid), phenicols (chloramphenicol), aminosides (gentamycin) => action on DNAgyrase and alteration of membrane permeability. |
| prophylaxis | -Chemoprevention and vaccination<br>-Rigorous hygiene before incubation<br>-Egg disinfection at all levels<br>-Rigorous hygiene of building, equipment and personnel<br>-Crawl space between strips |
| | |

### I.5.1.2.4 Avian Pasteurellosis

| Disease | | Avian Pasteurellosis |
|---|---|---|
| Synonymous names | | Avian cholera - Barbary disease - Coryza pasteurellica - Hemorrhagic septicemia - Pasteurella avisepticum - Pasteurella avium - Pasteurella avicida. |
| definition | | Bacterial, infectious, contagious, inoculable disease, Variable clinical expression |
| Epidemiology | | Bird species: All wild and domestic birds:<br>-Especially palmipeds (fattening birds: ducks + geese)<br>-Especially gallinaceous birds (turkey and chicken) Age:<br>Often acute in young people and chronic in adults:<br>-Duck and goose (4 to 5 weeks)<br>-chicken (+16 weeks) |
| Etiology (causes) | Pathogen | Pathogen:<br>*Bacillus or Coccobacillus, ovoid, G-, asporulate, capsulated, immobile,<br>*Pasteurellaceae family<br>**4 Genres:**<br>-Genus Pasteurella (pasteurellosis)<br>-G Actinobacillus (Actinobacillosis)<br>-G Haemophilus (Haemophilosis)<br>-G Avibacterium (tuberculosis)<br>***Species:**<br>-Pasteurella multocida: Avian cholera (especially palmipeds, chicken and turkey) => Pasteurellosis<br>-Pasteurella gallinarum: upper respiratory tract (chicken) => Pasteurellosis<br>-Pasteurella haemolytica: salpingitis, coryza, pneumonia, septicemia (especially chicken, rarely turkey, goose, quail, guinea fowl, pigeon).<br>**Characters:**<br>-Shelf life: 6 to 8 days in soil at room temperature - a few months in moist soil - 1 year in fresh clayey mud.<br>-Sensitive: sunlight (UV) - drying and heat (60°C for 10mn), disinfectants (formalin, phenol, soda, quaternary ammonium) |
| | Contaminating materials | Source of contamination:<br>Healthy carriers, sick subjects, cadavers (in a state of advanced putrefaction), cats, rats, mice, pigs<br>Contaminants:<br>All excretions (tears, droppings, saliva) except droppings (non-contaminating droppings) |
| | Penetration routes | Penetration routes:<br>oral, respiratory, ocular, transcutaneous (wound)<br>Transmission: Horizontal only (Direct and Indirect by water mainly).<br>Predisposing factors:Cold, humidity, stress of force-feeding, stress of long journeys, deficient diet, poor hygiene. |

| | | |
|---|---|---|
| | | Mortality rate:<br>-20%to70%=>epizootic form<br>-0%to10%=>enzootic form |
| | | |
| symptoms | | *Variable clinical expression:24h incubation experimentally<br>-F.suraigue:<br>Foudroyante = death in a few hours,<br>sadness, somnolence, inappetence, purplish cephalic appendages (crest + barbels) => death.<br>-F.aigue:<br>Very marked prostration, cyanosis of the crest and mucous membranes, hyperthermia => intense thirst, polypnoea, mucus in the nostrils and beak, liquid diarrhoea=> mucoid, greenish yellow to reddish=>.<br>foul-smelling "rainbow diarrhoea" => death within hours to days.<br>-F.chronic:<br>Coryza syndrome with the appearance of CKD (conjunctivitis, sinusitis, edema of the eyelids and barbels), |
| diagnosis | | D. clinical:<br>Symptoms and lesions<br>D.paraclinical:<br>-F. superacute and acute: blood + bone marrow + affected organs (liver, spleen, heart, intestine) + intestinal contents.<br>-F. chronic: nasal secretions + barb edema + joint lesions (abscesses) + affected organs (lungs) |
| treatment | | -Oral route: Antibiotics (sulfonamides, betalactamines, phenicols, quinolones, tetracyclines) + Vitamins (A - B - C). |
| prophylaxis | | -Chemoprevention and vaccination<br>-Improved breeding conditions<br>-Improved transport conditions<br>-Avoid cold and damp<br>-Eliminate sources of infection<br>Avoid the stress of force-feeding by providing polyvitamins |
| | | |

**I.5.1.2.5   Avian haemophilosis or infectious coryza**

| Disease | | Avian Haemophilosis |
|---|---|---|
| Synonymous names | | Infectious coryza - Infection with Avibacterium paragallinarum - <br> Avian hemophilosis paragallinarum (formerly) |
| definition | | Bacterial disease - Contagious-Respiratory tract infections <br> Upper respiratory). |
| Epidemiology | | Species: mainly chicken - pheasant - guinea fowl - pigeon. All ages: especially adults <br> Warm regions |
| Etiology (causes) | Pathogen | **Pathogen:Avibacterium paragallinarum** G- rod-shaped - mobile - non-spore-forming - 3 serotypes (A, B and C). <br> Characters: <br> -Sensitive: fragile in ambient environment inactivated at room temperature in 24h) <br> -Resistant to: cold (low temperatures) |
| | Contaminating materials | Source of contamination:Sick birds - healthy carriers. <br> Contaminants:Nasal and sinus discharge or exudates |
| | Penetration routes | **Entry routes:**Air (Respiratory) <br> **Horizontal transmission:** Direct (airborne) and indirect (drinking water - food - materials). <br> **Predisposing factors:** Environmental factors (high density - poor ventilation - wide temperature variations - stress). |
| | | |
| symptoms | | Incubation: 3 to 8 days - Duration: 1 to 2 weeks. <br> Clinical signs:Sluggishness - Decreased heart rate - Nasal discharge (serous then mucous) - Respiratory distress with rales - Birds shake their heads - Swollen head - Sneezing - Conjunctivitis - Swollen ridges. <br> Less frequent signs: swollen head with arthritis - diarrhea - egg-laying drop (10 to 40%). |
| diagnosis | | Clinical diagnosis:Respiratory signs (rales + respiratory distress + swollen head + infraorbital sinusitis + conjunctivitis). <br> Bacteriological diagnosis: <br> Detection from tracheal samples on blood agar or chocolate agar + biochemical characterization. PCR + serology (agglutination) |
| treatment | | Antibiotics:Macrolides (e.g. tylosin, tilmicosin); Aminosides; Sulfonamides; Tetracyclines |
| prophylaxis | | Vaccination; Biosecurity measures; Elimination of sick birds. |
| | | |

### I.5.1.3 Parasitic diseases

This group includes both internal and external parasites. Among internal parasites, the most frequently encountered are **coccidiosis**, which manifests itself in the presence of blood on the excrement, and at necropsy will show haemorrhages in the small intestine (intestinal coccidiosis) and in the caeca (caecal coccidiosis). The pathogenic agent of coccidiosis is a sporozoan protozoan of the Eimeridae family of the genus Eimeria, the species varying according to the avian species concerned. In hens, for example, we find E. acervulina, E. necatrix, E. maxima, E. brunetti, E. tenella, E. mitis, E. mitavi and E. praecox. Observable symptoms on the subject are:

> **Poor general condition of the hen;**

> **Inappetence and weight loss;**

> **Bloody diarrhea;**

Clinical diagnosis is made by observing bloody diarrhea and hemorrhagic lesions in the small intestine or cecum. If the disease is confirmed, anticoccidial treatment (e.g. toltrazuryl) should be administered.

As far as external parasites or ectoparasites of poultry are concerned, there are a multitude of them. They include

> **Fleas are biting, blood-sucking insects;**

> **Lice feed on integumentary debris;**

> **Red lice are hematophagous mites;**

> **Scabies, which are parasitic mites of the integuments;**

> **Ringworms are fungi that live on the integuments.**

The multiplication of ectoparasites in the building leads to intense discomfort for the hens and a drop in performance. The main signs of the presence of ectoparasites in the poultry house are:

> Chickens tend to scratch with their feet or search with their beaks for blood-sucking lice or ticks;

> Chickens will be constantly dust-bathing to remove lice, and feathers will look unattractive.

- ➢ We often see pecking;

- ➢ There will be weight loss and a drop in production performance;

- ➢ The lice will be visible inside the hen's feathers.

**I.5.1.4 Signs of mineral deficiency**

Chickens have well-defined mineral requirements, which vary according to their physiological state (maintenance, growth and production). For example, a growing hen will need 1% calcium (in %MS) in her daily ration, while the same hen in the laying phase will need around 4% Ca (%MS).

Mineral deficiency problems can be diagnosed at several levels:

- ➢ On eggshells: the shell becomes brittle, sometimes even discolored.

  <u>NB:</u> an excess of calcium in the ration can be seen in the white spots on the shell.

- ➢ On the sternum (the bone that separates the wishbone in two): a C- or S-shaped deformation of this bone can be observed;

Mineral deficiencies can be remedied by adding a mineral supplement to the water or to the ration, by formulating a ration balanced in minerals and adapted to the age and production of the hens.

**I.5.2 PROPHYLAXIS ON LAYING HEN FARMS**

Prophylaxis can be defined as all the techniques or measures to be taken to prevent the disease from taking hold or even spreading. Prophylactic measures therefore consist of complying with building construction standards, disinfecting the breeding site, setting up a control protocol for farm entrances and exits, and vaccinating animals against pathologies present in the breeding area.

**I.5.2.1 Impact of compliance with construction standards for livestock buildings**

There are certain standards for the construction of buildings for rearing laying hens that promote the well-being of the birds. The choice of site and type of construction will help or prevent the onset of certain pathologies. A site far from city noise will reduce animal stress. A well-ventilated building will prevent odors from developing, which also helps reduce stress and respiratory illnesses. In short, when building standards are respected, hens will be less stressed and therefore less sick.

## I.5.2.2 Disinfection and sanitary vacuum

Disinfecting livestock buildings involves washing them and using chemical substances to eradicate any germs present. The substances used to disinfect buildings are generally bactericides, virucides and fungicides. There are also molecules with a broad field of action that act on a wide range of microbes.

In addition to eliminating the microbes present, you'll also need to de-rat the site, as rats not only waste food but are also a major source of pathogens.

## I.5.2.3 Control of farm entrances and exits

To reduce the risk of spreading disease within the farm, all persons and vehicles entering the farm must follow a disinfection protocol. An SAS must be provided when the farm is built. The SAS is the place where farm personnel must clean up, change and wear work clothes before entering the rearing area. Every employee is required to wear boots and soak them in the foot bath located at the entrance to the rearing buildings. The SAS will not only be for employees, but all visitors to the farm must also pass through it, disinfect their hands and change into new clothes before entering the rearing area.

## I.5.2.4 Vaccination to prevent disease

Vaccination is a method of introducing a killed or attenuated pathogen, or even a fragment of the pathogen, into the animal's body, with the aim of provoking an immune response consisting of recognition of the pathogen by the immune system and production of memory antibodies that will react rapidly in the event of a real attack by this pathogen in the future.

Generally, the diseases vaccinated against are those that are recurrent in the breeding area. In Africa, the following diseases are vaccinated against:

- ➢ Marek's disease, vaccinated at the hatchery;
- ➢ New castle disease;
- ➢ Gumboro disease;
- ➢ Infectious bronchitis;

- ➢ Pasteurellosis;

- ➢ Collibacillosis;

- ➢ Smallpox;

- ➢ Avian encephalomyelitis;

- ➢ Corysa;

- ➢ Egg-laying drop syndrome or EDS.

There are several methods for administering these vaccines.

## I.5.2.4.1 Mass vaccination methods

As the name suggests, chickens are vaccinated en masse. There are two main types: vaccination in drinking water and vaccination by nebulization.

## I.5.2.4.1.1 Vaccination in drinking water

In this method, the vaccine is diluted in water and served to the animals. Precautionary measures are essential for successful vaccination.

First, the vaccine must be transported in a cooler to maintain the recommended storage temperature.

The hens to be vaccinated will be thirsty for two hours beforehand. It should be noted here that only healthy animals should be vaccinated. And this vaccination method is only applied to animals aged seven days or more.

The vaccine should be diluted in drinking water containing no disinfectants or detergents. Spring or well water without chemical treatment is ideal.

The vaccine solution should be served in clean troughs containing no traces of bleach or detergent. It should be noted that the vaccine must be consumed in its entirety within two hours of dilution, after which it will no longer be effective. You must therefore provide enough water troughs to satisfy all the animals, and calculate a quantity of water that can be consumed within two hours. This quantity generally represents 25 to 30% of the hens' daily consumption.

If we insist that the water used must not contain any traces of disinfectant, it's simply because the vaccine used is a live vaccine, and the principle of a live vaccine is that the micro-organisms at the origin of the disease are used in its manufacture. These microbes are attenuated, so they are no longer virulent and can therefore no longer induce disease. Administered to healthy subjects, they will be recognized by the animal's immune system, which will then produce memory antibodies capable of acting in the event of a real future attack. This means that using water containing disinfectant will destroy the vaccine and the animals will not be vaccinated. If you have any doubts about water quality, you can add a chlorine neutralizer to the water.

**I.5.2.4.1.2 Vaccination by nebulization or** spray

In this method, the vaccine is always diluted in water containing no disinfectant. Nebulization is a technique that involves spraying a vaccine into a closed building. Subjects are vaccinated by breathing in the air containing the vaccine. The droplets must be fine and homogeneous in size to induce good vaccination and avoid a vaccine reaction. For this purpose, there are several types of calibrated droplet devices:

- ➢ **Backpack nebulizers;**
- ➢ **Hand-held nebulizers;**
- ➢ **Cage carts.**

It's important to check that all types of device are in good working order before preparing vaccines, to avoid problems such as a flat battery or damaged nozzle once the vaccines have dissolved.

Maintenance is a key factor in nebulization, so there are several reasons why you need to keep your equipment in good working order:

- ➢ **Avoid clogging nozzles and the risk of delivering heterogeneous droplets;**
- ➢ **Avoid mold or other bacterial growth on the device that could contaminate animals;**
- ➢ **Avoid transmitting diseases from a previous band to the new band;**
- ➢ **Avoid using chemicals on the device that could destroy the vaccine;**
- ➢ **Avoid breakdowns and low battery problems during vaccination.**

After vaccination, always clean the device with spring water and store it in a clean, dust-free place.

### I.5.2.4.2 Individual vaccination methods

The special feature of these methods is that each subject receives an individual dose of vaccine. All subjects therefore receive the same dose. There are several routes to vaccination:

### I.5.2.4.2.1 Eye drops, beak drops and nasal drops

This method is applied using a calibrated dropper, generally a 30ml bottle capable of delivering 1,000 drops. The technique involves placing a drop of vaccine in the beak, nasal passage or eyeball. This method is used in primary vaccination for live vaccines such as Gumboro or New Castle. It helps develop local and general immunity, especially in view of the presence of the Harder's gland behind the third eyelid. It can be combined with an injectable vaccine.

### I.5.2.4.2.2 Subcutaneous and intramuscular injections

This method is generally used with inactivated vaccines in subjects over five weeks old. An automatic syringe is used, and intramuscular vaccines can be administered at the level of the wishbone muscle. The doses administered are generally between 0.3 and 0.5ml per subject.

### I.5.2.4.2.3 The winged transfixion

This method is used for the live fowlpox vaccine. A double-fluted needle is usually used to transfix the hen's wing membrane. This double needle is soaked in the vaccine solution before each transfixion.

# CHAPTER II COMPARATIVE STUDY OF TWO PONDEUS STRAINS: LOHMANN BROWN AND NORVOGEN BROWN

## II.1 Presentation of the farm

The farm where we carried out our studies is located in western Cameroon, in the Mifi department. This locality has an equatorial climate of the high-altitude Cameroonian type, characterized by a rainy season from mid-March to mid-November and a dry season from mid-November to mid-March.

The building used for chick start-up is an egg production building that will be modified to meet the needs of chicks in the start-up phase. This technique of modifying production buildings using plastic sheeting (photo 4) is very common in Cameroon, as very few farmers have a chick farm that meets chick rearing standards.

**Photo 4:**Building enclosed with plastic sheeting.

## II.2 Materials and methodology for setting up the nursery

## II.2.1 Materials used

For this comparative study of the two strains Norvogen and Lohmann in West Cameroon, layer production buildings were modified to create the microclimate needed by chicks in the start-up phase. These modifications include closing the side openings of the buildings with plastic sheeting to concentrate heat in the building. Gridding the building with bamboo and plywood to concentrate heating in specific areas, reduce the need for chicks to travel long distances, which wastes energy and leads to weight loss, and reduce competition by giving the impression of managing several small batches. Wood-burning ovens, still known as brulots, provided heat for the buildings.

## II.2.2 Room heating technology

To raise the ambient temperature in our rooms to a level recommended by the standard, depending on the age of the chicks, we used iron wood-burning ovens, also known as brulots, which are iron rods cut in two parts and reinforced with iron legs. Wood was used as fuel inside these burners.

At the coldest times of the day, the ovens were switched on to raise the temperature. When the sun was shining brightly, the ovens were switched off and the tarpaulins lifted to allow air to circulate and maintain the temperature at an adequate level. Photo 5 shows a wood-burning furnace.

**Photo 5:** wood-burning oven or brulot

## II.3- Chick rearing density

LOHMANN BROWN CLASSIC chicks supplied by SPC (Société des Provenderies du Cameroun) and NORVOGEN BROWN chicks supplied by SKAB Elevage were installed in buildings A and B at a density of 28 chicks/m$^2$ for the first 3 weeks. In the fourth week, some chicks were transferred to other buildings at a density of 16 chicks/m². At the end of week 10[i] the pullets were transferred to the other buildings for a final density of 8 hens per m . $^2$

## II.4 Feeders used

Since we didn't have any chick trays, we used the cardboard boxes in which the chicks arrived on their first day, which we took the trouble to trim to size (see photo 6) and used as first-age feeders up to twenty-one (21) days at a density of 50 chicks per feeder. We then switched to linear wooden feeders and age-appropriate siphon trays. From nine (09) weeks of age, we began to introduce third-age feeders for use during egg-laying.

**Photo 6:**Cardboard chick feeder

## II.5 Power supply

### II.5.1- Chick feed production

The feeds used were formulated with VIAL France 2.5% Premix and other nutrient-rich ingredients, while respecting the needs of the chicks according to their age and weight. Thus we had:

> **Energy sources:** corn, wheat bran;

> **Protein sources:** VIAL2.5% premix and soy, cotton and peanut meals;

> **Mineral sources:** calcined bone powder and marine shell;

> **Vitamins and antitoxins.**

So we made our own food, using the grinder-mixer, which grinds and mixes ingredients according to standards.

**II.5.2-** Feeding **system**

We applied a split-feeding system, feeding the chicks three times a day to stimulate feed consumption:

> **At 7:30 in the morning**, the feed troughs were cleaned and 2/4 of the total feed was served;

> **At 12:00** we returned to stir the feed and serve 1/4 of the total feed;

> **At around 2:30 or 3pm**, the leftovers from the day's 1/4 feed were served.

For the first three weeks, water containing a multi-vitamin (Maxi Layer) and a hepatoprotectant (Heparenol) was served every morning before 6 a.m. and renewed at 2 p.m..

## II.6 Zootechnical performance of chicks and pullets

### II.6.1 Summary table of zootechnical performance

| AGE | Live weight (in g) | | Feed consumption (g/p/d) | | Total feed consumption (in g) | | Weekly mortality | |
|---|---|---|---|---|---|---|---|---|
| | lohmann | norvogen | lohmann | norvogen | lohmann | Norvogen | % Lohmann | % Norvogen |
| 01 | 60,4 | **65,5** | 9,1 | **9,29** | **63,7** | 65,03 | 1,15 | 1,6 |
| 02 | 97,07 | **87,3** | 16,7 | **15,25** | **180,6** | 171,78 | 0,31 | 0,16 |
| 03 | 154,12 | **158,7** | 21,21 | **23,96** | **329,07** | 339,5 | 0,067 | 0,1 |
| 04 | 224,5 | **205** | 29,73 | **23,63** | **537,18** | 504,91 | 0,061 | 0,11 |
| 05 | 295 | **272,3** | 26,97 | **28,85** | **725,97** | 706,86 | 0,077 | 0,1 |
| 06 | 439 | **353,5** | 39,37 | **37,66** | **1001,56** | 970,48 | 0,061 | 0,39 |
| 07 | 527,5 | **446,4** | 45,32 | **44,47** | **1318,8** | 1281,77 | 0,46 | 0,28 |
| 08 | 595 | **538,1** | 52,91 | **42,6** | **1689,17** | 1579,97 | 0,31 | 0,21 |
| 09 | 679 | **608,9** | 45,36 | **42,56** | **2006,69** | 1877,89 | 0,041 | 0,21 |
| 10 | 783,6 | **679,6** | 56,6 | **54,05** | **2402,89** | 2256,24 | 0,49 | 0,18 |
| 11 | 863 | **802,3** | 51,8 | **57,7** | **2765,49** | 2660,14 | 0,23 | 0,35 |
| 12 | 960 | **871** | 77,6 | **52,41** | **3308,69** | 3027,01 | 0,02 | 0,1 |
| 13 | 1087 | **945,7** | 59,3 | **60,91** | **3723,79** | 3453,38 | 0,041 | 0,51 |
| 14 | 1142 | **1037,2** | 66,89 | **59,35** | **4191,81** | 3847,62 | 00 | 0,06 |
| 15 | 1196 | **1114,5** | 60,2 | **65,07** | **4613,21** | 4303,11 | 0,005 | 0,06 |
| 16 | 1252 | **1213,5** | 63,9 | **67,98** | **5060,51** | 4778,97 | 0,02 | 0,028 |
| 17 | 1379 | **1273,4** | 71,24 | **71,04** | **5559,19** | **5276,25** | 0,159 | 0,028 |
| 18 | 1549 | **1279,9** | 80,98 | **63,92** | **6126,05** | **5723,69** | 0,01 | 0,076 |
| 19 | 1484,43 | **1394,8** | 92,9 | **76,02** | **6776,35** | **6255,83** | 0,051 | 0,042 |

<u>**COMMENT**</u>

Subjects' food consumption is recorded each evening, based on the number of 50kg bags served. Weights are recorded every weekend. 200 hens are weighed using a five-gram (5g) precision electronic balance.

We can see that feed consumption and live weights for both strains of hen are increasing. We also note that in some weeks feed intake was low. The factors affecting feed intake and consequently weight gain are:

- Stress arising from handling subjects;

- Pathologies;

- Food quality;

- Endogenous factors such as strain (some strains consume more feed);

- Ambient temperature (the warmer it is, the less the hens consume the feed).

At the end of week 19, the cumulative mortality rate of the Lohmann brown strain (3.56%) was lower than that of the Norvogen brown strain (4.59%). This indicates greater viability of the lohmann brown.

**II.6.2 Food consumption curve for the two strains**

The hens' feed consumption was assessed at the end of each week, and averaged for each hen per day. Figure 1 below compares the daily feed consumption of the LOHMANN BROWN strain with that of the NORVOGEN BROWN strain.

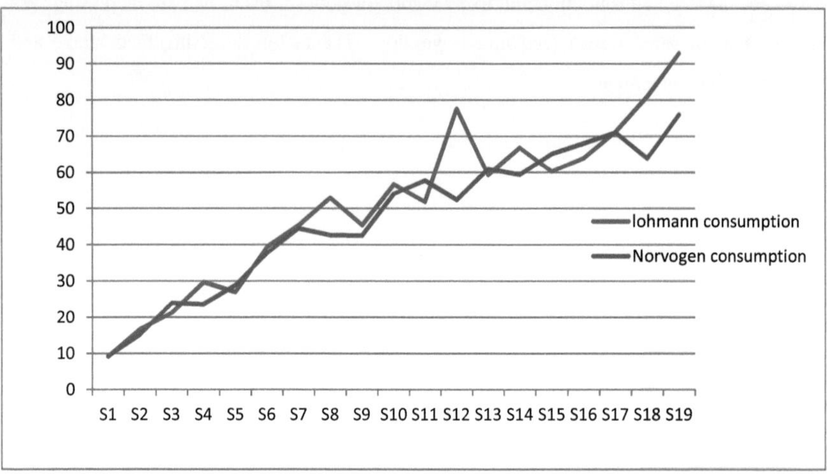


**FIGURE 1**: Comparative food consumption curve

**COMMENT**

In general, consumption of both strains increased. We also note that the Lohmann brown consumption curve was above that of Norvogen brown most of the time (1 to 2 weeks; 6 to 11 weeks; 12 to 15 weeks and at the end 17 to 19 weeks), demonstrating that the Lohmann brown strain consumed more feed than the Norvogen brown strain during the chick rearing phase.

## II.6.3 Live weight trends for the two strains

Figure 2 compares the live weights of Lohmann brown and Norvogen brown.

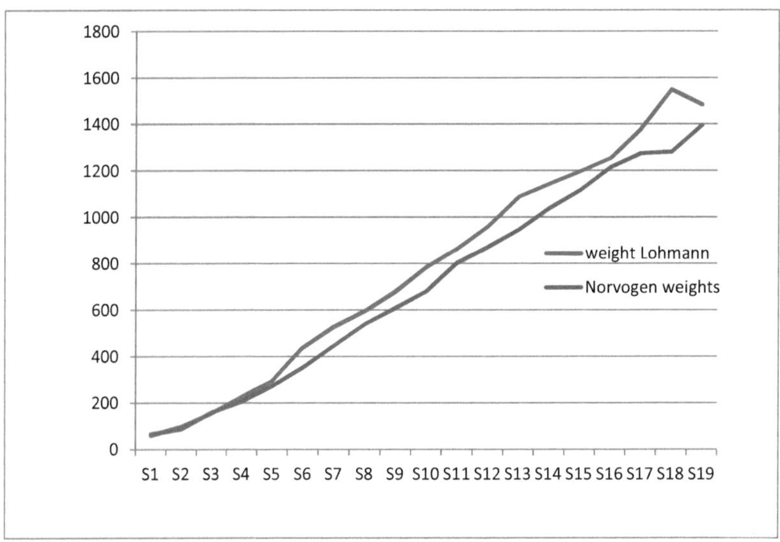


**FIGURE 2** Live weight curve

## COMMENT

In figure 2 above, we can see that the Lohmann brown strain's live weight curve is above that of Norvogen brown throughout the rearing period from week 1 to week 19. This means that Lohmann brown had a better weight than Norvogen brown. Figure 1 already showed a higher feed intake in Lohmann brown. The higher feed intake of the Lohmann brown resulted in a higher weight in figure 2, so it's clear that the closer the chicks' feed intake is to the norm, the better their weight will be.

# CONCLUSION

The main aim of this document is to describe how laying hen chicks are reared in Cameroon and to provide information on the zootechnical performance of two strains of laying hens. This is to help table egg producers choose the strains that will be most beneficial to their production objectives. With this in mind, we compared the growth performance of two laying strains (Lohmann brown and Norvogen brown) reared in West Cameroon. The results showed that feed consumption and weekly weights of the Lohmann strain were higher than those of the Norvogen strain throughout the rearing period. At 19 weeks, the mortality rate of the Lohmann strain reached 3.5% (i.e. a viability of 96.5%), while that of the Norvogen strain reached 4.4% (the norm for mortality in the rearing phase being between 2% and 3%). These parameters are important in determining a strain's adaptability to a given environment, enabling growers to choose the most suitable strain for their production objectives. For greater certainty in the choice of strain to be used in West Cameroon, further work will be carried out on the egg-laying performance of these two strains in the rearing conditions of this region.

# BIBLIOGRAPHY

- FICHE-UCSC-ZOOT-Lignes guides l'élevage familial de poules pondeuses final vision17012015

- Breeding guide for commercial layers-NOVOgen BROWN.

- Classic Lohmann Brown Breeding Guide

- Mansouri S., 2009. comparaison des performances zootechniques entre deux souches de poules pondeuses (ISA Brown et LHOMANN Brown) au niveau de l'O.R.A.V.I.O . Mémoire présenté en vue de l'obtention du diplôme d'ingénieur agronome à l'université de Ibn Khaldoun de Tiaret. 58p

# I want morebooks!

Buy your books fast and straightforward online - at one of world's fastest growing online book stores! Environmentally sound due to Print-on-Demand technologies.

Buy your books online at
**www.morebooks.shop**

Kaufen Sie Ihre Bücher schnell und unkompliziert online – auf einer der am schnellsten wachsenden Buchhandelsplattformen weltweit! Dank Print-On-Demand umwelt- und ressourcenschonend produzi ert.

Bücher schneller online kaufen
**www.morebooks.shop**

Printed by Books on Demand GmbH, Norderstedt / Germany